Dedication

This book is dedicated to my mother and father, Daisy and Frank. You both taught me how to understand the world better through imagination and storytelling. You told the best stories. The values and lessons you have communicated through them are not forgotten. Rest in Peace.

TABLE OF CONTENTS

Introduction

As a mathematics teacher of 22 years, I've noticed several students struggle with determining which of the basic operations to use when solving word problems. Many learners fail to connect the key words and phrases that give them the clues as to which number operation to perform. Without the ability to make the connections between key words and the appropriate number operation, solving word problems can become frustrating for the learner. As a result, the learner may tend to avoid the challenge of unraveling word problems all together.

As quoted by Daniel Wallace, "A storyteller makes up things to help other people." Storytelling has been used as a practice for instruction for thousands of years. Kosa (2008) claimed, "Stories transmit values, engage the imagination, and foster community." Furthermore, applicable storytelling draws the learners' attention and addresses each type of learner in becoming engaged with the lesson. Since the storyteller appeals to the students' dominant senses of communication, learning is more apt to occur.

In the words of Robert McAfee Brown, "Storytelling is the most powerful way to put ideas into the world today." This book tells a story of the four basic number operations. The story personifies the operations as medical doctors: **Dr. Addny Addition, Dr. Sidney Subtraction, Dr. Molly Multiplication and Dr. David Division**. Each of the four doctors tells their story of when and how they operate on numbers through experiences of their own; listening to key words and phrases.

Dr. Addny Addition

Dr. Add

Hello, my name is Doctor Addny Addition. My friends call me **Doctor Add** for short. I am one of the four doctors that perform operations on numbers. You'll meet my three partners soon. Being a number operations doctor is very important and unique. You see, the numbers that come into our operating rooms only need one special operation to solve problems. Follow me to the next page and I'll tell you all about my operating room, **The Addition Room**.

The numbers that visit my operating room all want to be stitched together; that way, they make one combined number.

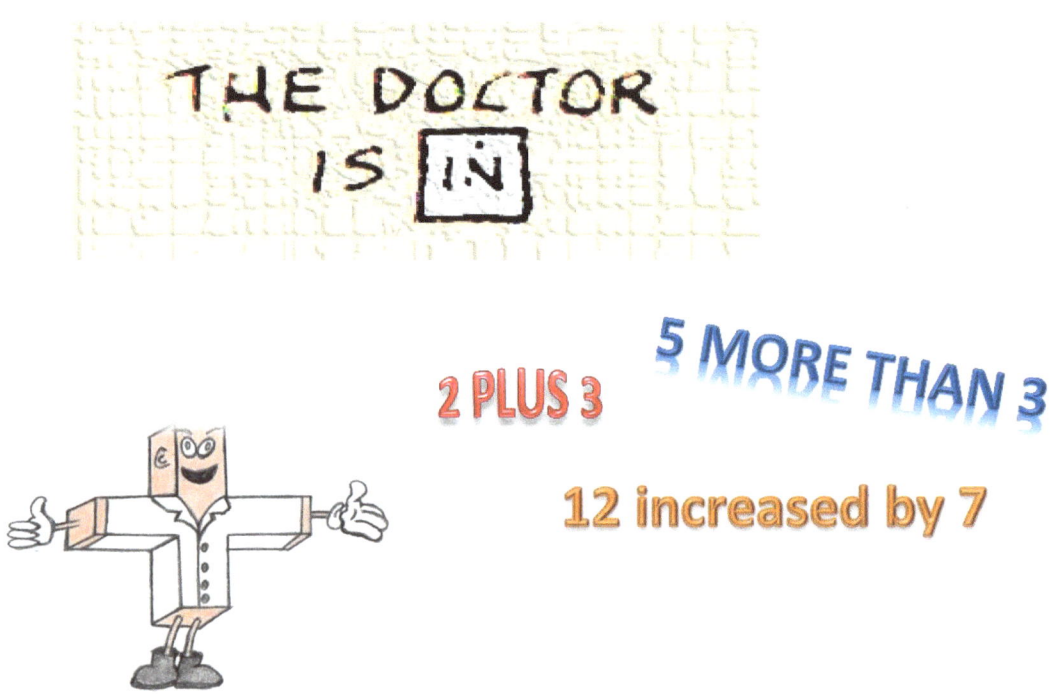

The numbers in my operating room use lots of key words and phrases to describe their symptoms. The key words and phrases listed below tell me to perform the **operation of addition** on these numbers:

Why just the other day, the numbers 5 and 2 came into my office asking to be **joined** together. The number 5 recently proposed, "I want to be **increased by** 2!" The number 2 was flattered by the offer. You see, the number 2 was excited that their **sum** would make her five **more than** her current value. What number doesn't want to be **more than** what they are?

I quickly assured the both of them they had come to the right doctor. Since my job as Dr. Add is to stitch numbers **together**, I was glad to perform the **operation of addition** for them. I happily led them into my operating room .

I'm proud to say the addition operation was a success! After routine surgery, the number 7 was the result of the happy union. As a matter of fact, the number 7 was so pleased with the operation, it celebrated by having a party.

As one of the most requested doctors of number operations, you can imagine the volume of numbers I stitch together every day. I have lots of stories to share with you. But, for the sake of time, I'll only tell a few:

Patient Requests	Operation of Addition	Result
"I want to be 3 more than 7"	3 + 7 =	10
Four and One wanted me to combine them together.	4 + 1 =	5
Intercom: "Dr. Add, 8 and 9 want their values totaled."	8 + 9 =	17
Four and Two asked to be joined together on Valentine's day.	4 + 2 =	6
Once three numbers came in wanting to know their combined sum. The numbers were 2, 5, and 9.	2 + 5 + 9 =	16
One family of four numbers came into my office and asked, "What is our value in all?" The numbers were 2, 4, 6, and 8.	2 + 4 + 6 + 8 =	20

Remember:

Think of me, Dr. Add, when there are numbers to stitch,

"Altogether" or *"all in",* doesn't matter which,

Dr. Add is there to help you **combine,**

Any of the numbers on the number line.

When two was **increased by** one, they made a family of three,

It couldn't have happened without me,

The **operation of addition**, that's my niche,

Adding numbers together stitch by stitch!

Yours Truly, Dr. Add

Oh my! Dr. Add is falling behind schedule because so many of his number patients need the **addition operation** performed on them. He needs you to help him with the numbers in his operating room. Please perform the **operation of addition** on the numbers in the following problems. **Circle ALL key words and phrases** that tell you to perform the **operation of addition**.

1. Linda read 3 books and Bob read 7 books. How many books did Linda and Bob read in all?

2. Valerie and Tammy went to the mall to get a breakfast smoothie. Valerie's order was 6 dollars and Tammy's order was 5 dollars. Tammy agreed to pay for both smoothies. What is the total Tammy has to pay?

3. Maria had 10 pair of shoes in her closet. Her mother gave her 2 more for her birthday. How many pair of shoes did Maria have altogether?

4. The water gage was at 2 inches on Tuesday. After the rain on Wednesday, it increased by 4 inches. How many combined inches of rain fell?

5. Last year, Lucy pulled the football away before Charlie could punt on 6 occasions. This year Lucy has only pulled the football away on 3 occasions. Including last year and this year, what is the total number of times Lucy pulled the football away from Charlie?

Dr. Sidney Subtraction

Dr. Subtract

So, how did you like performing the operation of addition with Dr. Add? Dr. Add is known for making math fun! Not to mention, bankers love him.

Now that you've gotten your feet wet working problems with Dr. Add, it's time to expand your practice with another number operation. In the world of number unions, sometimes numbers want their stitches removed or they just want to be reduced in size. That's where I, Dr. Subtract, can help. You can say I perform the opposite operation of Dr. Add. Don't get me wrong, my partner, Dr. Add, performs a wonderful service when he stitches numbers together. Why, many number tales would end in sad tragedy without the contributions of Dr. Add. But, after some numbers are stitched, they sometimes begin to feel like they have taken on too much. The desire to lose, what we number operation doctors call "number fat", consumes them. They simply have to lose weight or dwindle down for one reason or another.

All of my number patients require the **operation of subtraction**. You can say that performing the **operation of subtraction** is like undoing stitched numbers or **reducing** number fat. Recall the numbers 5 and 2 were stitched together by Dr. Add. Their sum resulted in the number 7. Now the number 7 wants to be **reduced by** 3. Some numbers are never satisfied! Who am I to question the choices of numbers? So, I'll perform the **operation of subtraction** for the number 7 later for you to see.

Well, here I am. I've just arrived at my office and my operating room is prepped and ready for number surgery. Come on in and watch me operate on numbers.

The Office of
Sidney Subtraction, ND (Number Doctor)
Hours: 7:00 am to 5:00 pm
Specializing in the
Operation of Subtraction

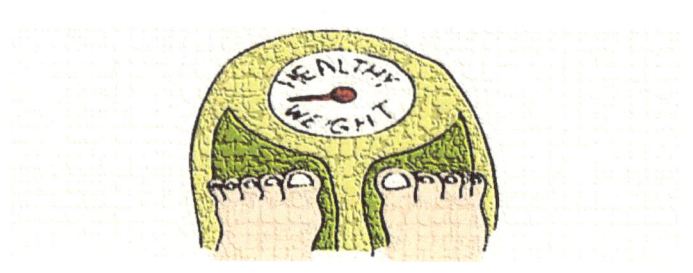

Before	After
Size 18	Size 6

WEIGHT LOST EQUALS:

THE **DIFFERENCE** BETWEEN 18 AND 6

18 MINUS "?" = 6

18 IS **HOW** MUCH MORE THAN 6?

Like Dr. Add's number patients, my number patients describe their symptoms with key words and phrases too. As a number operation doctor, it is very important to pay close attention to key words and phrases. The following key words and phrases tell me to perform the **operation of subtraction**.

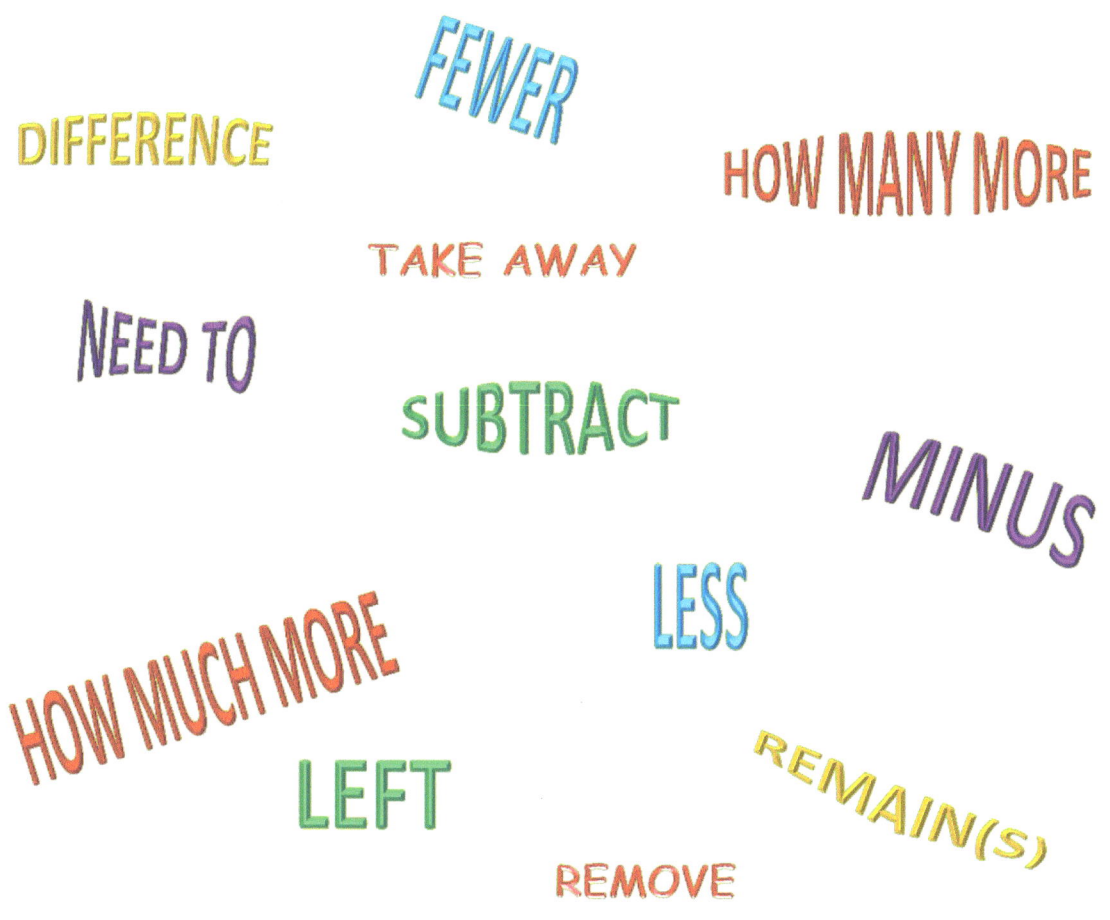

The first number patient on my schedule is the number 7. Remember, the number 7 requested to be **reduced by** the number 3. Let's step into my operating room where the number 7 is waiting .

Paying attention to key words and phrases guided my procedure. The phrase "**reduced by**" told me that the number 7 wanted me to **subtract** the number 3 from it. As you can see, the result of the **subtraction operation** is the number 4; *3* **less than** *7* or 7 less 3.

Join me for a few more of my operations. The practice will help you to understand my use and purpose with numbers. If you get stumped, remember to check out the **key words**

and phrases that tell you to perform the **operation of subtraction (see page 18)**. Now, please join me as I use the **operation of subtraction** to help the number patients on my morning schedule.

Dr. Subtract's Morning Schedule

Number Patient Requests	Perform the Operation of Subtraction	Result
8:00 am – The number 8 and the number 3 request the difference in their value.	8 – 3 =	5
8:30 am – The number 12 would like to know how much more weight he has to lose before he is a number 9.	12 – 9 =	3
9:00 am – The number 6 wanted to remove his stitches and asked to take away the number 3.	6 – 3 =	3
9:30 am – The number 11 wants to know what will remain after she drops the number 6.	11 – 6 =	5
10:00 am – The number 10 wants to be 8 fewer than he is now.	10 – 8 =	2
10:30 am – The number 25 is losing weight and wants to know how many more units before she becomes the number 14.	25 – 14 =	11
11:00 am - Mr. Fifteen wants to know what life would be like minus his girlfriend, Ms. Eleven.	15 – 11 =	4
11:30 am to 12:30 pm - LUNCH	--------	--------

DON'T Ignore:

If it's the **difference** the numbers ask

Dr. Subtract is the one for the task

She is more than a star in the cast

Of number doctors in surgical mask

Keep her in mind, when you need to **subtract**

Pay attention to key words, that's a fact

Words like "**fewer**" and "**less**" tell us to act

And use the subtraction operation with tact

SINCERELY, DR. SIDNEY SUBTRACTION

Dr. Subtract really enjoyed you tagging along with her as she helped her number patients this morning. Her afternoon workload is much more demanding. She is asking for your assistance. Please perform the **operation of subtraction** on the numbers in the following cases. **Circle ALL key words and phrases** that tell you to perform the **operation of subtraction**. *Please show your work*.

1. Felecia bought a candy apple for $5.00 and a hamburger for $8.00. How much less did the candy apple cost?

2. Brad spent $3.00 on cotton candy. He gave the cashier a $20 bill. How much money does Brad have remaining?

3. John had 15 baseball cards. His friend Michael traded him three Emoji cards for three of his baseball cards.

How many baseball cards did John have left after the trade?

4. Keri and Brittani are both girl scouts. This year Keri sold 25 cases of cookies and Brittani sold 21 cases. How many more cases of cookies did Keri sell than Brittani?

5. Ceasar started the checkers game with 12 pieces. After five minutes into the game he was down to 8 pieces. How many less pieces does Ceasar have after five minutes of play than when he began?

Dr. Molly Multiplication

DR. MULTIPLY

Wow, as you have seen, number patients can be very finicky about the operations they choose to have performed on them! As their conditions change, the reasons for their operations differ from one number patient to the next. The number patients' demands determine the type of number operation done to them. Luckily, the number operation doctors are here to assist the number patients with their wishes regardless of how choosy they are.

Let me introduce myself. My name is Dr. Molly Multiplication. I'm better known as Dr. Multiply. Number patients come to me when they need more than one or multiple stitches. You can say I'm like Dr. Add on the fast track. My number patients usually want to stitch the same number value together many times.

For instance, I once had a number patient that wanted to become 6 times its current value of 5. In other words, he wanted to be the total sum of 5 + 5 + 5 + 5 + 5 + 5. Well, I could have stitched 5 more of the number 5 to its value. That would certainly create a total of 6 of the number 5 stitched together making a sum of 30.

I know you are probably thinking this looks like a number patient request for Dr. Add. If you recall, Dr. Add specializes in stitching numbers together. But, when number patients request to have the same number added together multiple times, that is when I'm called upon to perform the operation of multiplication. You see 5 + 5 + 5 + 5 + 5 + 5 is the same as multiplying 5 times 6.

The result for both procedures is 30. Besides, it is less stressful on the number patient to undergo one operation of multiplication than many operations of addition. There is also less chance for making mistakes. Come into my waiting room and observe the patients' requests 🚪.

Office of Multiplication Operations

All Surgeries Performed by

Dr. Multiply, ND Specialist

10 TIMES 3

THE PRODUCT OF 4 AND 2

TWICE 6

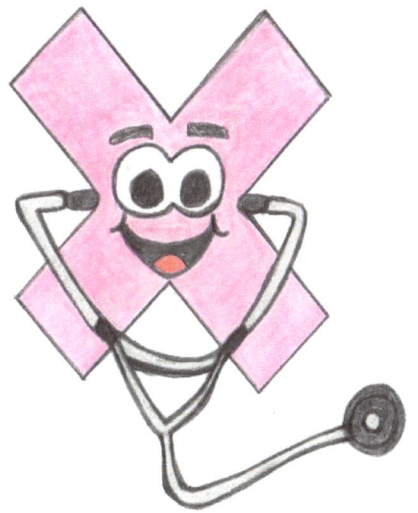

As you can see, my number patients describe their symptoms in unique ways just like the number patients that rely upon Dr. Add and Dr. Subtract. That is why it is very important to listen carefully to the number patient's warning signs. By paying close attention to their key words and phrases, we number operation doctors know exactly which operation to perform. Thank heavens my number patients do not rely upon as many key words and phrases to explain their operation needs. The number patient's that visit my office usually express their concerns with the following key words and phrases:

When I was on vacation in the state of Washington last year I ran across a wonderful fruit market. One of the salespersons was selling assorted apples. He cut a piece of one of the apples so that I could taste it.

The flavor was like music to my mouth. I thought, "These apples would be perfect for making my world famous apple punch!" Without hesitation I decided to bring as many of the apples home as my luggage could carry. The apples were packaged six to a bag. My luggage could only fit 4 times the size of the bag of apples. By nature, I performed the operation of multiplication to determine the total number of apples I had purchased as my apple punch recipe calls for 12 large apples. Please enter into my operating room and allow me to give you a glimpse of the **multiplication operation** I performed .

+ + + = ?

REMEMBER: Since we are stitching the **same number** of apples together **multiple times**, I stepped in to perform the **operation of multiplication ONLY once**:

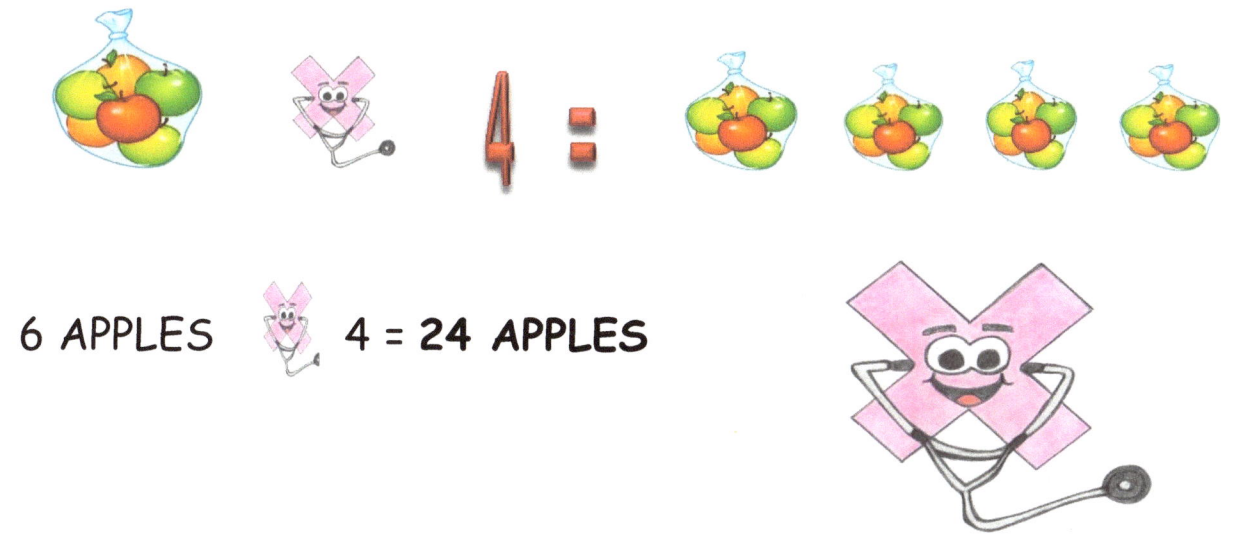

6 APPLES ✕ 4 = **24 APPLES**

     As it turned out, I had bought more than enough apples to make my apple punch. Being the number doctor that I am, I thanked the salesperson for his service and promised to visit him again when I returned to Washington State. I also gave a glowing recommendation to my number patients. You know the old saying, "An apple a day, keeps the doctor away."

     Unfortunately, not all of my patients like apples as much as I do. This means there are always number patients in need of the **operation of multiplication.** Come along and explore a few of my cases with me. **Focus on the key words and phrases.**

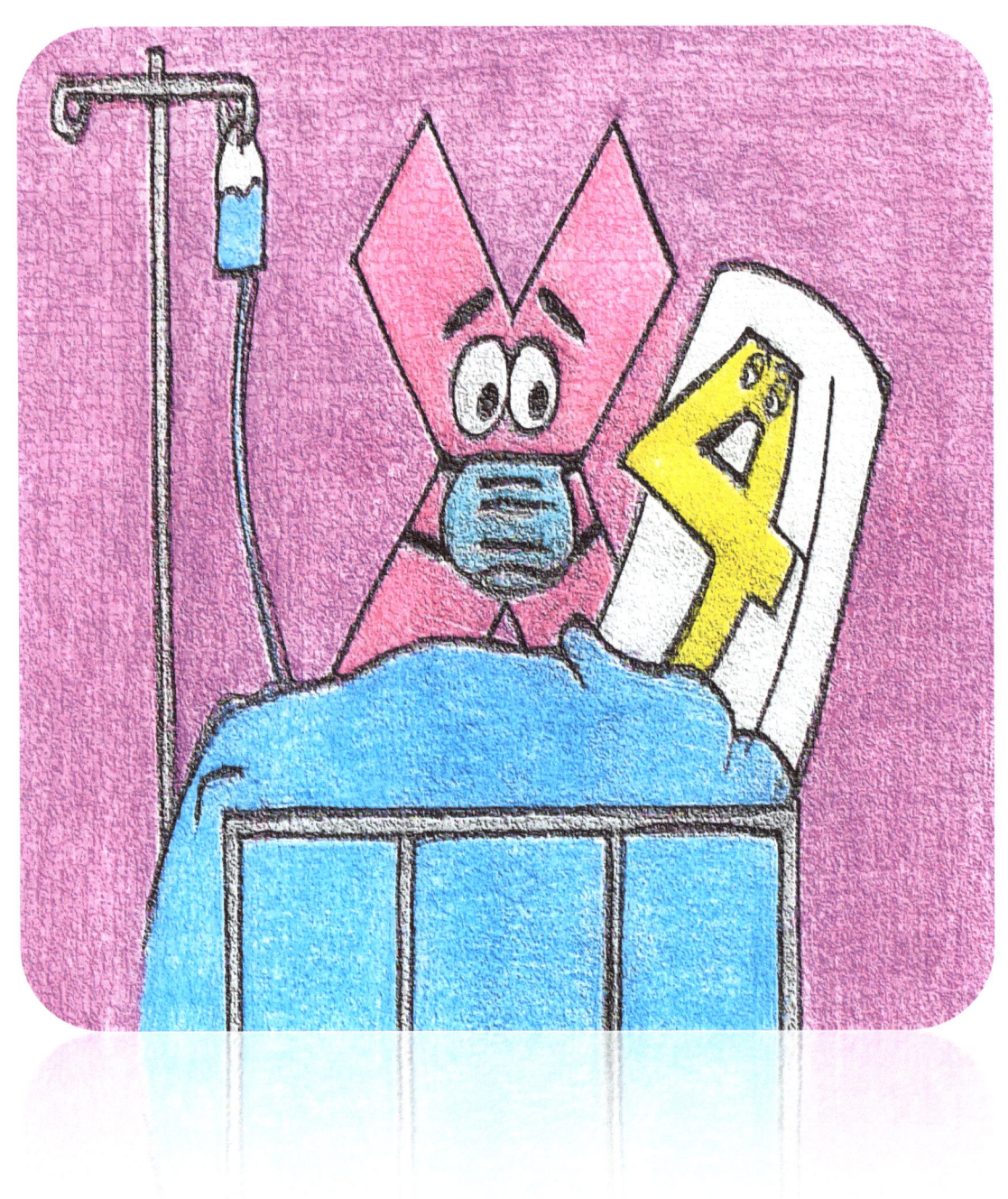

Number Patient Cases for Dr. Multiply		
Number Patient Requests	Operation of Multiplication	Outcome
The number 3 needs the product of itself and 4	**** **** **** 4 x 3 =	12
The number 6 wants to know the outcome of twice its value	****** ****** 6 x 2 =	12
The number 7 is asking to become 4 times her value	******* ******* ******* ******* 7 x 4 =	28
There is a request to have 8 multiplied by 6	******** ******** ******** ******** ******** ******** 8 x 6 =	48

Don't FORGET.........

When addition is repeated

And the number is the same

Think of me, Dr. Multiply

That's just my game

Some call it skip counting

Many **times** if you're able

But you can always find **products**

Using my **multiplication** table

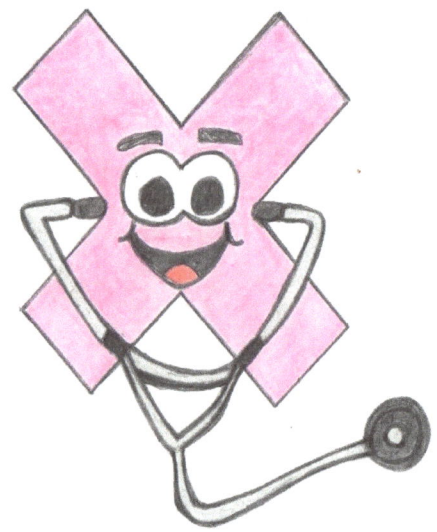

REPEATINGLY YOURS, DR. MULTIPLY

Dr. Multiply wants you to know she had a blast with you as you visited with her and her number patients. She hopes you continue to pay close attention to the number patients' key words and phrases. Always keep in mind the key words and phrases are your clue to performing the right number operation. It is now your turn to perform the **operation of multiplication.** Remember, **multiplication is repeated addition of the same number**. For further help, Dr. Multiply has placed a multiplication table chart in the back of this book to assist you with your operations of multiplication. Enjoy your practice! **Make sure you circle all key words and phrases.**

1. If Brandy drinks 8 glasses of water each day, how many times does she drink water in one week? Note: One week is 7 days.

2. What is the product of 4 times 3?

3. John has 8 transformer cars. Jeffrey has twice as many as John. How many transformer cars does Jeffrey have?

4. Over time Carl saw his number of chicks multiplied by 4. If Carl began with 3 chicks, how many chicks does he have now?

5. Asia won only 3 tennis games. Her sister, Ty, won 3 times as many games as Asia. How many games did Ty win?

Dr. David Division

Dr. Divide

Oh my, you sure have had loads of practice in the number operating rooms. I was told by my three number doctor (ND) partners, Dr. Add, Dr. Subtract and Dr. Multiply, that you have really been on a give and take journey of number operations. Your practice is molding you into experts at the number operation craft. You have learned to listen to number patients through key words and phrases. I have to say, "I am proud of the work you have done!" My number patients have a few more key words and phrases to share with you. Are you ready to learn more from my number patients?

I am the last number doctor (ND) you will meet in our partnership of number operations. My name is Dr. David Division, better known as Dr. Divide. I undo multiple stitches of the same number by performing the operation of division. In other words, my job is to determine how many times one number value can be taken away from another number value. The operation of division is like repeated subtraction of the same number or skip counting backwards. It is the reverse operation to multiplication.

For example, three boys had a hobby of collecting odd rocks. One weekend the boys went camping in the Grand Canyon. While walking throughout the canyon the boys found 12 rare rocks among them.

By the end of the trip, the three boys sought to split the 12 rocks evenly among them. As a way to divide them evenly, the boys thought to give the rocks away in sets of three until the rocks were all gone. That is one rock to each boy until the rocks were gone:

12 rocks – 3 rocks – 3 rocks -3 rocks -3 rocks = 0 rocks.

The boys quickly noticed 3 rocks could be subtracted 4 times from 12 rocks. This meant each boy would get to keep 4 rocks to add to his collection. The boys were proud of their discoveries and planned to share them with their school mates during "*Show and Tell*".

Coincidently, I am very good friends with one of the boys' family. One night when I was over for a bar-be-que, the boy told me the story about how the three of them used repeated subtraction to split the 12 rocks evenly among them. I spoke up and said, "You know, I specialize in repeated subtraction. But, instead of performing multiple operations of subtraction, I perform one **operation of division**.

	Boy 1	Boy 2	Boy 3
First Round of Subtracting 3 Second Round of Subtracting 3 Third Round of Subtracting 3 Fourth Round of Subtracting 3	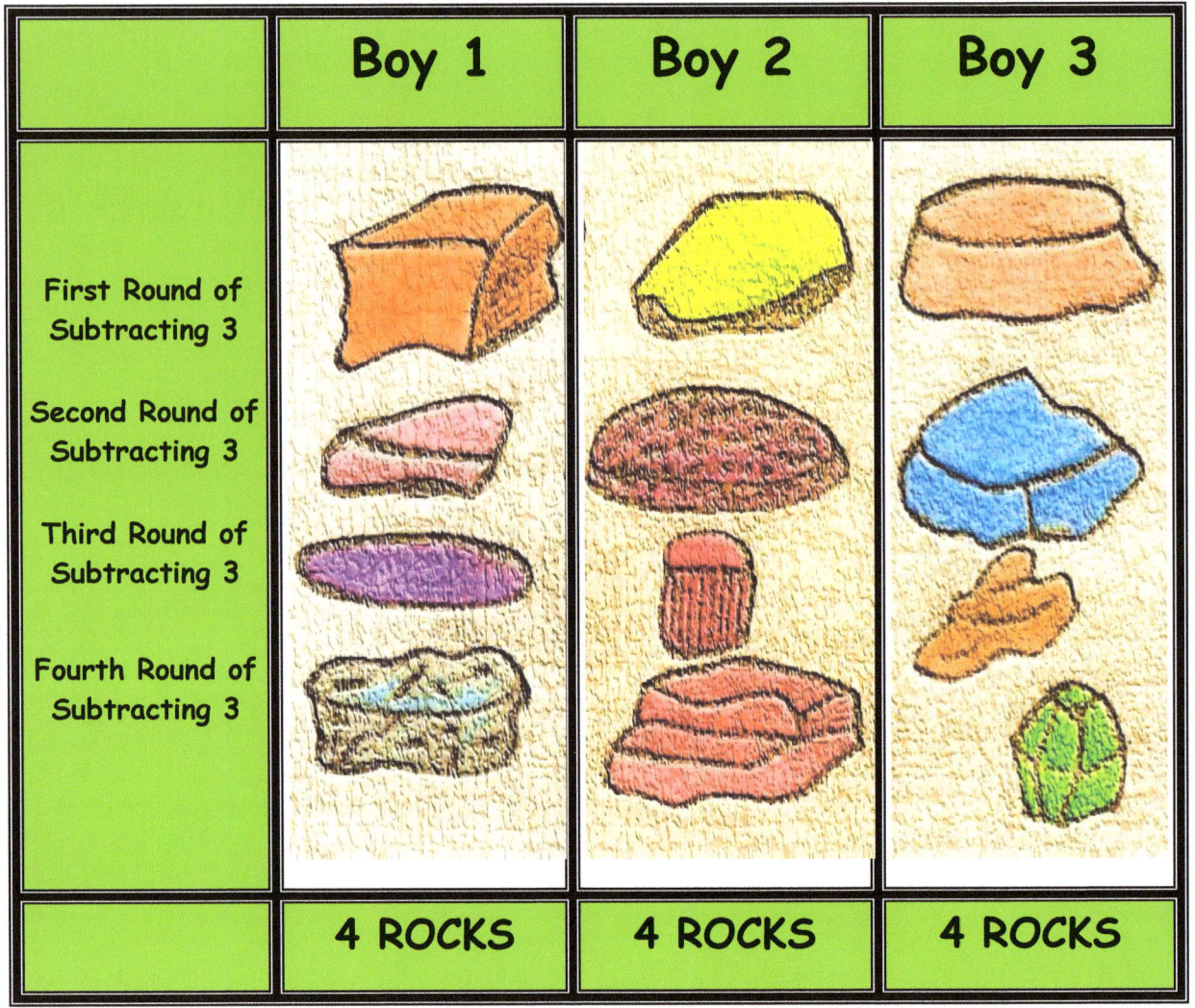		
	4 ROCKS	4 ROCKS	4 ROCKS

Notice, the number of times the boys subtracted 3 from 12 until they arrived at 0 is the same as dividing 12 by 3. You are invited to take a walk with me as I teach you about the operation of division. Get ready to get blown away! Come in to my office and meet some of my number patients.

The Office of DR. David DIVIDE

Specializing in the

Operation of Division

(Formally known as *backwards skip counting or repeated subtraction*)

15 DIVIDED BY 5

The Quotient of 6 and 2

DIVIDE 12 INTO GROUPS OF 3

My number patients are screaming for help. Do you hear the key words and phrases they are using to describe their conditions? The number patients that seek the **operation of division** routinely use these key words and phrases.

Last weekend, I attended a dinner party with the other number operation doctors. Together we made a party of four. So, we sat at a table with only four place settings.

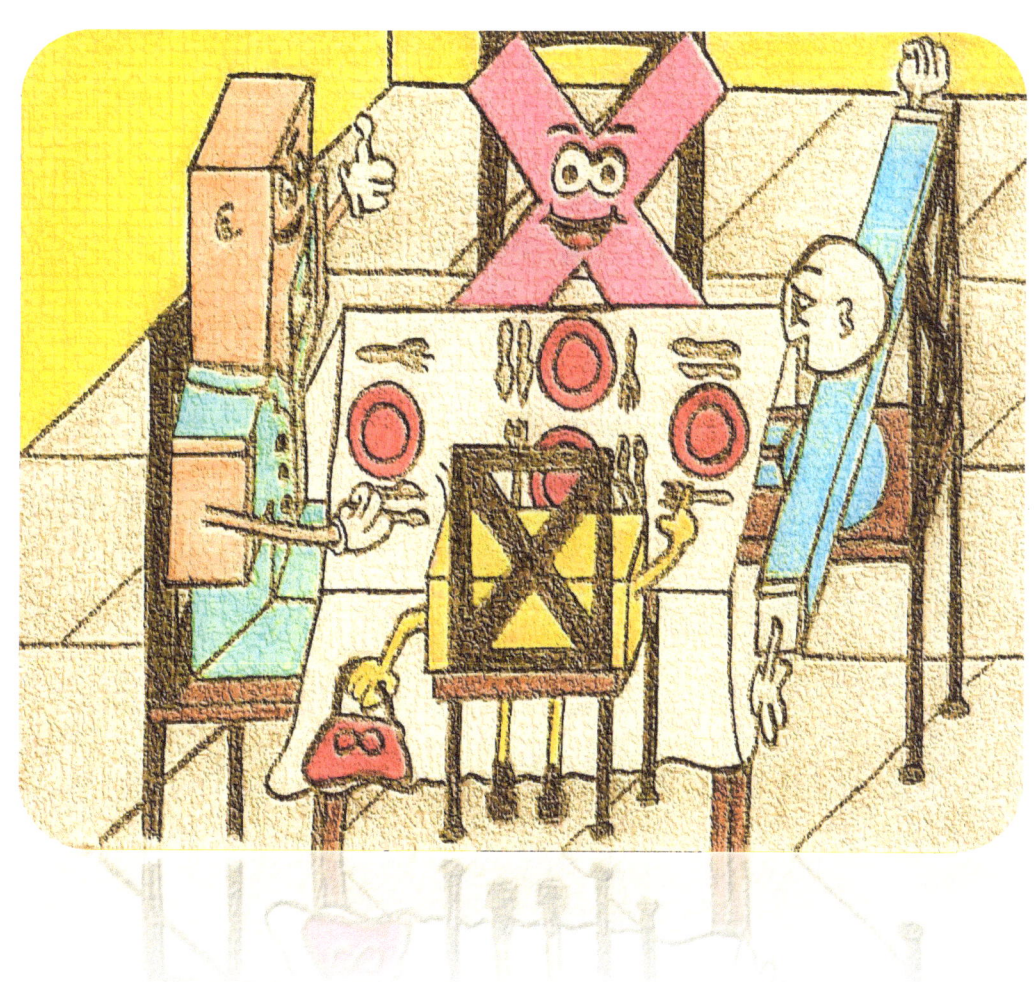

I remember, the host and hostess served spaghetti and meatballs with garlic bread for the main course. Not only was the sauce the best I have ever had, the meatballs were to die for. Each table of four was given 16 meatballs on a platter of spaghetti and savory sauce. As a matter of fact, the meatballs were so good, no one at our table wanted to part with their even share. So, to keep the meatball portions equal among the four of us, I decided to perform the **operation of division**.

Imagine taking away 4 meatballs at a time, one for each ND, from the 16 original meatballs. Continue with this process until the meatballs are all divided among the four number doctors. That is: 16 – 4 – 4 – 4 – 4 = 0; 4 sets of 4 meatballs (🍖🍖 🍖🍖). Well that's exactly the plan of the **operation of division.** But instead of asking number patients to go through repeated operations of subtraction, I perform one **operation of division**. Number patients seeking the operation of division want to know how many of one number can be taken away from another number or how many times can one number be divided by another number. Come into my surgery room and let's explore the operation of division I performed for my partners .

    As I mentioned before, we could have relied on Dr. Subtract to determine how many times the number four could be subtracted from 16.

---

16 Meatballs – 4 meatballs – 4 meatballs – 4 meatballs – 4 meatballs = 0 meatballs

Note: The number 4 can be taken away from 16 four times. Therefore,

---

But, because Dr. subtract would have had to perform four subtraction operations of the same value, I offered to save time by performing only one **operation of division**.

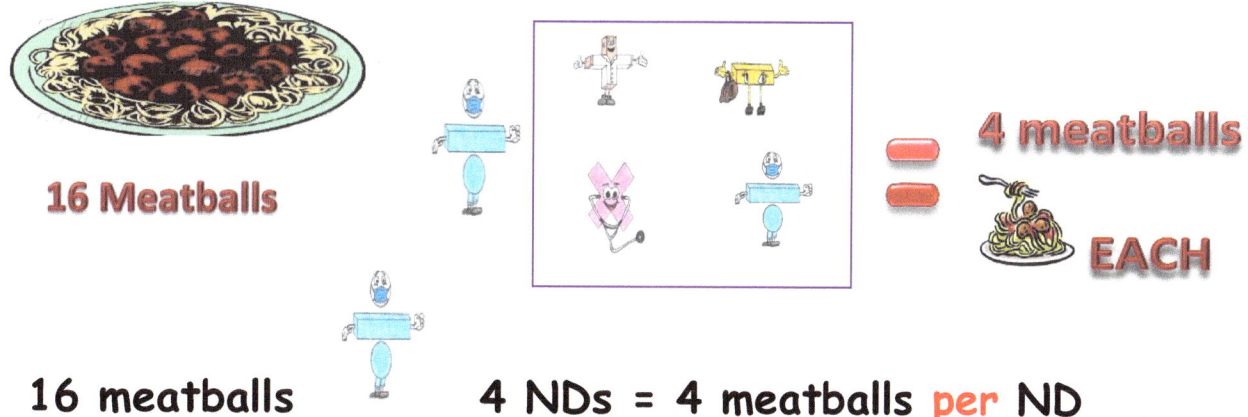

16 Meatballs

16 meatballs   4 NDs = 4 meatballs per ND

That's right! We each received four meatballs per plate. Yes indeed, those meatballs were yummy to my tummy and I didn't leave a crumb! That reminds me, I must send a thank you card to the host and hostess for serving such a lovely dinner.

Come along, we have a busy schedule today. As my helpers, I'm going to need you to pay close attention to key words and phrases. Keep in mind, the key words and phrases tell us rather or not our number patients are in the right number doctor's office. Let's take a look at my operation of division schedule from last week. Notice how I remembered to listen to the number patients.

DR. DIVIDES WEEKLY OPERATION SCHEDULE

Office Closed on Wednesdays and Fridays

Week Of ___LAST WEEK_____

Patient Condition	Operation of Division	Result
Monday @ 8:00 am- The number 10 wants the quotient between it and the number 5.	$10 \div 5 =$	2
Monday @ 10:00 am - The number 2 wants the number of times it goes into the number 14.	$14 \div 2 =$	7
Tuesday @ 9:00 am – 20 marbles want to be divided evenly among 5 marble collectors. (Note to self: How many marbles per collector)	$20 \div 5 =$	4
Tuesday @ 2:00 pm – The number 8 wants half its value.	$8 \div 2 =$	4
Thursday @ 9:00 am – The OLD WOMAN WHO LIVES IN A SHOE needs to divide her 20 children into groups of 4.	$20 \div 4 =$	5

Keep In Mind….

Rather small or rather quite wide

Number patients call on Dr. Divide

to evenly split values or equally provide

Matching sets of numbers hidden inside

If it's the quotient you need

Or just an even break

Dr. Divide will make sure

You get a fair shake

EQUALLY YOURS, DR. DAVID DIVIDE

CONGRATULATIONS, You made it through your final training with the number operation doctors. Dr. Divide has told me how smart you all are and can't wait to have you visit with him again. However, right now Dr. Divide needs your help. He has been called out of town on a family emergency and cannot help his number patients this week. He needs you to handle his case load below. He has confidence in your ability. There is only one thing. Dr. Divide asked me to remind you to ALWAYS pay attention to the **key words and phrases**. Remember to listen to the number patients. **Circle ALL key words and phrases that tell you to perform the operation of division.**

1. Tyler had 50 crayons and 5 boxes. How many crayons should he pack in each box so each guest gets the same number of crayons?

2. Leslie had 9 baseball caps. He wanted to divide them evenly among his three brothers. Can you tell Leslie the number of times 3 goes into 9?

3. Frank's school hung 30 posters in the auditorium. They distributed the posters in 5 even rows. How many posters are in each row?

4. Daisy bought 72 oranges. She divided them equally into 9 baskets. How many oranges did Daisy put in each basket?

5. In the Art room, each table seats 5 people. How many tables will a group of 15 people need?

VOCABULARY

Addition: The number operation of finding the total or sum among two or more numbers

Assorted: Consisting of a number of different kinds

Confidence: Trust or faith in a person or thing

Consumes: To fully engage

Current: Happening now; here and now

Difference: The amount by which one number is greater or less than another number

Division: The number operation of determining how many times one number is contained in another number; the opposite of multiplication

Dwindle: To slowly become smaller

Fast Track: The quickest and most direct way to achieve a goal

Finicky: Hard to please; choosy

Flattered: To be complimented too much and often dishonestly.

Glimpse: A quick view or look

Hesitation: To be slow to act

Key Word: A word used as a clue to provide support or direction

Multiple: A number that can be evenly divided by another number

Multiplication: The number operation that involves adding a number to itself a certain number of times

ND: Number Doctor

Number Operation: A process or an action performed on numbers according to special key words and phrases

Operation: A surgical way for treating an injury, a complaint, a defect, or a fault

Opposite: Altogether different

Phrase: A meaningful arrangement of words that provides support or direction

Pitch: Talk designed to motivate

Practice:	The act or process of doing something; performance or action
Routine:	Normal; Regular
Salutations:	A polite saying or greeting
Skip Count:	Counting forwards or backwards by a number other than 1
Stitch:	To fasten together with staples or thread
Subtraction:	The number operation of finding the difference between two numbers
Sum:	An amount found as a result of adding numbers
Symptom:	A specific sign or clue of the presence of something else
Tragedy:	An unhappy ending
Undergo:	To experience something
Union:	Joined together
Value:	A given or worked out number amount
Workload	The amount of work assigned to or expected from a worker in a given time period

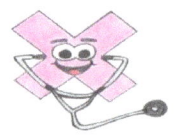

Multiplication Table
(Up to 12)

Zero's	One's	Two's	Three's
0 x 0 = 0	1 x 0 = 0	2 x 0 = 0	3 x 0 = 0
0 x 1 = 0	1 x 1 = 1	2 x 1 = 2	3 x 1 = 3
0 x 2 = 0	1 x 2 = 2	2 x 2 = 4	3 x 2 = 6
0 x 3 = 0	1 x 3 = 3	2 x 3 = 6	3 x 3 = 9
0 x 4 = 0	1 x 4 = 4	2 x 4 = 8	3 x 4 = 12
0 x 5 = 0	1 x 5 = 5	2 x 5 = 10	3 x 5 = 15
0 x 6 = 0	1 x 6 = 6	2 x 6 = 12	3 x 6 = 18
0 x 7 = 0	1 x 7 = 7	2 x 7 = 14	3 x 7 = 21
0 x 8 = 0	1 x 8 = 8	2 x 8 = 16	3 x 8 = 24
0 x 9 = 0	1 x 9 = 9	2 x 9 = 18	3 x 9 = 27
0 x 10 = 0	1 x 10 = 10	2 x 10 = 20	3 x 10 = 30
0 x 11 = 0	1 x 11 = 11	2 x 11 = 22	3 x 11 = 33
0 x 12 = 0	1 x 12 = 12	2 x 12 = 24	3 x 12 = 36

Four's	Five's	Six's	Seven's
4 x 0 = 0	5 x 0 = 0	6 x 0 = 0	7 x 0 = 0
4 x 1 = 4	5 x 1 = 5	6 x 1 = 6	7 x 1 = 7
4 x 2 = 8	5 x 2 = 10	6 x 2 = 12	7 x 2 = 14
4 x 3 = 12	5 x 3 = 15	6 x 3 = 18	7 x 3 = 21
4 x 4 = 16	5 x 4 = 20	6 x 4 = 24	7 x 4 = 28
4 x 5 = 20	5 x 5 = 25	6 x 5 = 30	7 x 5 = 35
4 x 6 = 24	5 x 6 = 30	6 x 6 = 36	7 x 6 = 42
4 x 7 = 28	5 x 7 = 35	6 x 7 = 42	7 x 7 = 49
4 x 8 = 32	5 x 8 = 40	6 x 8 = 48	7 x 8 = 56
4 x 9 = 36	5 x 9 = 45	6 x 9 = 54	7 x 9 = 63
4 x 10 = 40	5 x 10 = 50	6 x 10 = 60	7 x 10 = 70
4 x 11 = 44	5 x 11 = 55	6 x 11 = 66	7 x 11 = 77
4 x 12 = 48	5 x 12 = 60	6 x 12 = 72	7 x 12 = 84

Eight's	Nine's	Ten's	Eleven's
8 x 0 = 0	9 x 0 = 0	10 x 0 = 0	11 x 0 = 0
8 x 1 = 8	9 x 1 = 9	10 x 1 = 10	11 x 1 = 11
8 x 2 = 16	9 x 2 = 18	10 x 2 = 20	11 x 2 = 22
8 x 3 = 24	9 x 3 = 27	10 x 3 = 30	11 x 3 = 33
8 x 4 = 32	9 x 4 = 36	10 x 4 = 40	11 x 4 = 44
8 x 5 = 40	9 x 5 = 45	10 x 5 = 50	11 x 5 = 55
8 x 6 = 48	9 x 6 = 54	10 x 6 = 60	11 x 6 = 66
8 x 7 = 56	9 x 7 = 63	10 x 7 = 70	11 x 7 = 77
8 x 8 = 64	9 x 8 = 72	10 x 8 = 80	11 x 8 = 88
8 x 9 = 72	9 x 9 = 81	10 x 9 = 90	11 x 9 = 99
8 x 10 = 80	9 x 10 = 90	10 x 10 = 100	11 x 10 = 110
8 x 11 = 88	9 x 11 = 99	10 x 11 = 110	11 x 11 = 121
8 x 12 = 96	9 x 12 = 108	10 x 12 = 120	11 x 12 = 132

Twelve's
12 x 0 = 0
12 x 1 = 12
12 x 2 = 24
12 x 3 = 36
12 x 4 = 48
12 x 5 = 60
12 x 6 = 72
12 x 7 = 84
12 x 8 = 96
12 x 9 = 108
12 x 10 = 120
12 x 11 = 132
12 x 12 =144

www.ingramcontent.com/pod-product-compliance
Lightning Source LLC
Chambersburg PA
CBHW050802180526
45159CB00004B/1519